Sebastian Wagner

Declared Distances and Safety Areas

TORA, TODA, ASDA, LDA and

GRIN Verlag

Bibliografische Information der Deutschen Nationalbibliothek:

Die Deutsche Bibliothek verzeichnet diese Publikation in der Deutschen Nationalbibliografie; detaillierte bibliografische Daten sind im Internet über http://dnb.d-nb.de/ abrufbar.

Imprint:

Druck und Bindung: Books on Demand GmbH, Norderstedt Germany
ISBN: 978-3-656-46750-2

This book at GRIN:

http://www.grin.com/en/e-book/230404/declared-distances-and-safety-areas

Declared Distances and Safety Areas

TORA, TODA, ASDA, LDA and RESA

SEBASTIAN WAGNER

Air Traffic Control / Air Traffic Management

AVIMA 12

June 2013

Table of Contents

1 Introduction

In a conventional air transport system, airports and aircraft are highly dependent on each other in providing service for their customers and passengers. Especially advances in the engine and airframe technology forced airports to plan, design, and construct facilities necessary to ensure that they were not left behind in full participation in a high growth industry. Nowadays international airports are progressively faced with huge aircraft with greatly increased wingspans carrying almost 900 people. These new circumstances affect airport layouts eminently. Beside the configuration of cargo and passenger terminals, geometric and structural design of pavements, including taxiways, exists and aprons, number and orientation of runways, the **length of the runway** is a main factor which has to match the improved parameters (*Ashford/Mumayiz/Wright 2011, p. 74 ff.*). Nevertheless, international airports operate with a wide range of present-day aircraft which have quite different requirements of the runway. At this point it is up to the airport owner to declare distances on the runway which are available for satisfying the takeoff runs and landings of airplanes. We know, the longer the runway the higher the costs for sure; but with the provision of declared distances the more and larger aircraft can be attracted by the airport especially for small airports with relatively short runways.

It is exactly the mentioned runway which plays a central role in the entire airport facility system. The runway has always been the most crucial point of all maneuvering areas where all takeoff and landing movements take place. It is proven that takeoff and landing phases are the most critical parts of all stages of a flight. They are demanding the highest skills from the crews and placing huge stresses and strains on the aircraft itself (*Croucher 2007, p. 12-1*). Therefore it is no wonder that events in which an aircraft veers off or overruns the runway surface during either takeoff or landing, so called **runway excursions**, are the most likeliest emergency situations. According to the Air Transport Safety Survey, runway excursions are the **most common type of accident** reported annually both in the European region and elsewhere in the world. 106 excursions with commercial and executive aircraft operations worldwide were recorded in 2012. That means there are at least two runway excursions each week worldwide (*Air Transport News 2013, p. 13*).

ICAO Aircraft Accident/Incident Data Reports (ADREP) have shown that aircraft undershooting or overrunning the runway during landing or takeoff suffer significant damage. To minimize such damage, it is considered necessary to provide structural measures or additional areas at and beyond the end of the runway or even surrounding the landing strip. Subsequently, provisions of **declared distances and safety areas** make an important contribution to preclude potentially hazards to aircraft and to protect passengers and people on the ground in case of an aircraft veers off or overruns the runway.

In the first two chapters I am going to explain the various kinds and differences of declared distances by explaining the relevant background and basic terminology with the help of illustrations. In the following I will briefly identify application areas of declared distances and how or rather why they have to be published for flight crews. Finally the paper concludes with a small insight in the field of safety areas in particular the Runway End Safety Area.

2 Background and Terminology

In applying declared distances and safety areas in airport design, it is helpful to understand the relationship between aircraft certification, aircraft operating rules, and airport data and design.

During the entire takeoff procedure the complex aircraft systems are running under full throttle. In this phase of flight various technical malfunctions, the recognition of a significant abnormality or an ATC instruction to stop the aircraft can occur that is forcing the pilot to abort the departure until reaching the calculated **takeoff decision speed (V_1)**. This speed is computed by the pilots in advance based on the airport data/design and can be understood as a "Go or No Go Decision". At speeds below V_1, the aircraft may be able to stop before the end of the runway. Once V_1 has been exceeded, the takeoff must be continued even if an emergency occurs.

Additionally aircraft certification provides the aircraft's **performance distances**. They are based on the performance speeds/data of each individual aircraft established by the manufacturer and confirmed during certification testing for varying climatologically conditions, operating weights and much more. Typical performance distances are:

- **Takeoff Run** – The distance to accelerate from break release to lift off
- **Takeoff Distance** – The distance to accelerate from break release past lift off to start for takeoff climb
- **Accelerate-Stop Distance** – The distance to accelerate from break release to V1 and then decelerate to a stop
- **Landing Distance** – The distance from the threshold to complete the approach, touchdown, and decelerate the approach.

All distances are published by the manufacture and internalized by the type-rated pilots. These distances, based on specific parameters mentioned above, are given and cannot be changed by the crew. To all distances a certain safety factor is included (*FAA 2012, p. 57*).

Now we are going into detail and I will define and explain the basic terms and acronyms which are from evident importance to understand the system of safety areas and declared distances. Basic definitions are given in Annex 14 and Aerodrome Design Manual (Doc 9157), both published by the International Civil Aviation Organization (ICAO).

As mentioned in the introduction, the length of the runway is from upmost importance and influences the aerodromes performance data decisively. The field length is determined not only on the basis of the aircraft's design but also by the safety regulations established by the responsible bodies of individual member countries of the ICAO. The actual runway length should always be adequate to meet the operational requirements of the airplanes for which the runway is intended. Furthermore, the length should be not less than the longest length determined by applying the corrections for local conditions to the operations and performance characteristics of the relevant aircraft (*ICAO Annex 14, 3.1.6*).

In some cases airport operators plan to extend the dimensions of the runway. Reasons can be variegated. For instance they try to achieve higher economic benefits by attracting new airlines with larger aircraft. Often operators have an agenda to reach socio-economic goals or their intention is to improve the level of service; Or they simply try to improve safety. Whatever the reasons are, **runway extension decisions can be very costly and complex** (*LPS Aviation 2007, p. 5*).

In particular out of safety perspectives it can be absolutely worthwhile to increase the length of the runway. If you have a closer look on the accident statistics you will see that overrunning the end of the runway during takeoff and landing are one of the most common types of aircraft accidents. In this regard a useful alternative of extending the actual length of runway can be the so called **Stopway (SWY)**. The SWY constructed at the end of runways as emergency space to slowly stop planes that overrun the runway to slowly stop a plane on a rejected takeoff.
According to ICAO Annex 14 SWY is defined as follows: "*A defined rectangular area on the ground at the end of take-off run available prepared as a* ***suitable area in which an aircraft can be stopped in the case of an abandoned take-off***."

IMAGE 1: SIMPLIFIED ILLUSTRATION OF A STOPWAY (*OWN ILLUSTRATION*)

Nevertheless, if a SWY is established it must fulfill various conditions. According to the regulation of the respective aviation authority it must be able to support the weight of the largest aircraft able to use and shall be constructed with sufficient strength to support a decelerating aircraft in all weather conditions. Additionally, the aerodrome operator designates and commits to maintaining the area as stopway, including the appropriate lighting and marking shown in the image below (*FAA 2000, p. 1*).

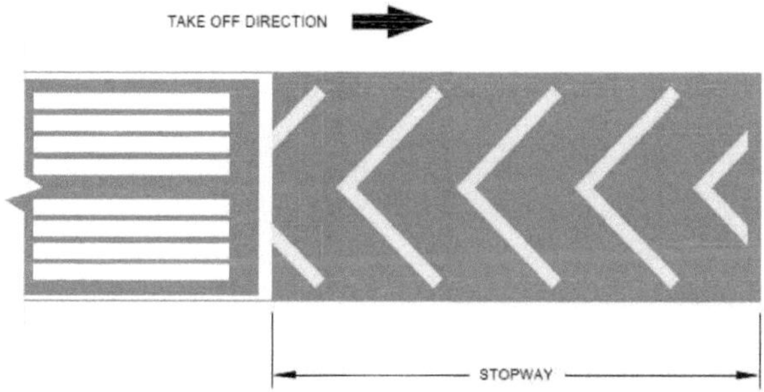

IMAGE 2: EXAMPLE OF THE PROPER MARKING OF A STOPWAY (*FAA 2012, P. 93*)

A second option to increase the length parameters of a runway is the establishment of a **Clearway (CWY)**. A CWY is a nearly flat part on the ground or water and is not a paved part of the runway like the SWY is. It should be understood as a safety area beyond the runway end, which allows an airplane to fly over in low altitude after takeoff without the threat of getting endangered by any major obstacles. Generally it does not extend beyond the airport boundary. Again ICAO Annex 14 provides an official definition: *"A defined rectangular area on the ground or water under the control of the appropriate authority selected or prepared as a* ***suitable area over which an aeroplane may make a portion of its initial climb*** *to a specified height."* The image below depicts a simplified figure of the installation of a clearway. Aside from that the provision of a SWY is not mandatory to establish a CWY.

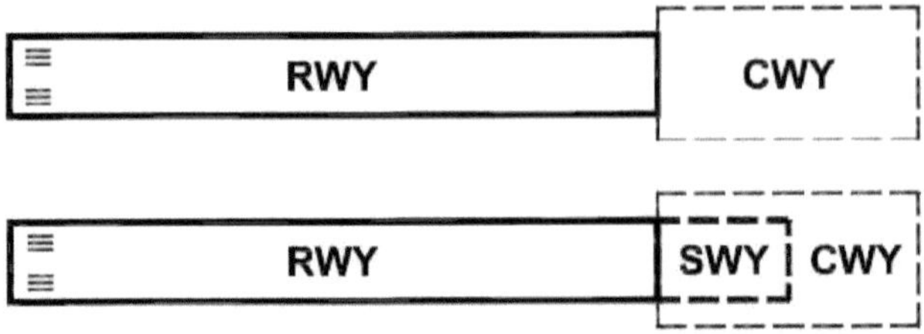

IMAGE 3: SIMPLIFIED ILLUSTRATION OF A CLEARWAY (*OWN ILLUSTRATION*)

Basically for large airports the application of SWY and CWY does not make any sense in most cases because they do have a relative long runway anyway. However, for example for smaller regional airports with limited financial resources, which are in addition short of space on the airside area, the establishment of a SWY/CWY can be a useful option. As a result airport planners can benefit from both alternatives. It is frequently advantageous to establish a clearway because it **saves full strength runway without penalizing the operational usage**. More precisely a clearway increases the allowable aircraft operating takeoff weight without increasing the length of the runway (*Ashford/Mumayiz/Wright 2011, p. 90; FAA 2012, p. 65*). Unfortunately most ICAO recommendations concerning SWY and CWY published in Annex 14 are ambiguous and are not uniformly applied worldwide. Whenever a crew is in doubt about the quality of a SWY/CWY, consideration in performance optimization tools may be inappropriate (*Bos 2013, p. 1*). Nevertheless, whenever a SWY/CWY is provided, the stopway length and the declared distances shall be provided in the airport/facility directory and in the aeronautical information publication (AIP) for international airports for each operational direction (*ICAO Doc 9157, 3.4.2*). A practical example will follow in the appendix.

Before finally entering the main subject of declared distances, I need to make a short explanation of the term "**Displaced Threshold**". The landing threshold marks the beginning of that portion of the runway usable for landing and is usually located at the beginning of the runway. However, displacement of the landing threshold may be required when an object that obstructs the airspace required for landing aircraft is beyond the airport owner's power to remove, relocate or lower. Thresholds may also be displaced for environmental considerations, such as noise abatement, or to provide additional safety space. Hence, the displacement of the threshold shortens the usable runway length for landing, while not affecting the distance of the runway available for departing aircraft (*FAA 2012, p. 52*). Whenever a displaced threshold is encountered it must be identified, classified, and documented similarly to a runway end.

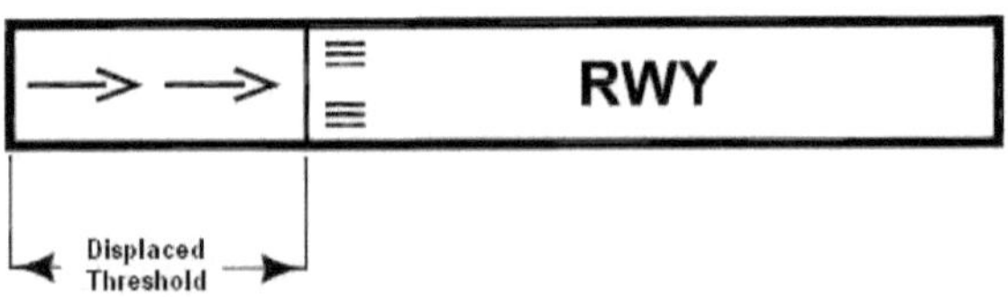

IMAGE 4: SIMPLIFIED ILLUSTRATION OF A DISPLACED THRESHOLD (*OWN ILLUSTRATION*)

3 Declared Distances

The introduction of stopways, clearways and the use of displaced thresholds on runways has created a need for accurate information regarding the various physical distances available and suitable of takeoff and landing of turbine powered airplanes. These information are from upmost importance for flight crews who need to calculate the available distances for each runway direction during pre-planning (*ICAO Doc 9157, 3.4*). For these purposes, the term "declared distances" is used with the following distances associated with a particular runway. To say it in other words, declared distances simply **represent the maximum distances available** for meeting aircraft's (rejected) takeoff and landing performance requirements, **which exceed the official physical limits of the runway** (*FAA 2012, p. 56*).

3.1 Take-Off Run Available (TORA)

TORA describes the length of the runway declared available und suitable for the **ground run of an aircraft during take-off**.

Neither the stopway nor the clearway is taken into consideration for the calculation of TORA. It is simply the full runway distance available for take-off run at which the aircraft accelerates until the end of the (physical strengthened part of) runway. Additionally, if a clearway exists and it begins prior to the runway end, the TORA ends at the beginning of the clearway.

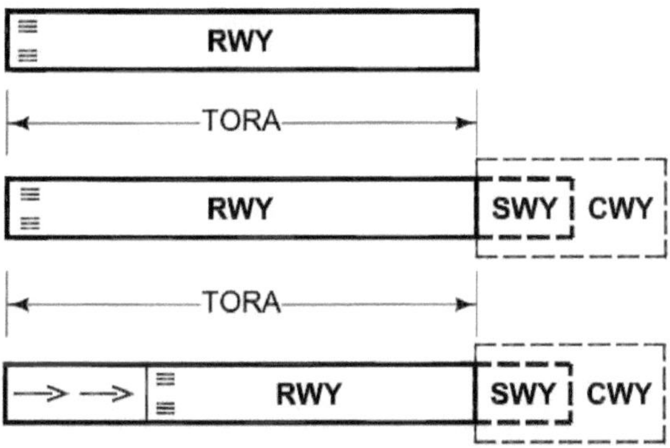

IMAGE 5: SIMPLIFIED ILLUSTRATION OF TORA (*OWN ILLUSTRATION*)

In case of the existence of a displaced threshold the pilot will be able to start the take-off run before the suitable landing zone of runway. In this case the displaced threshold increases the available runway length to conduct the takeoff run successfully. All information are concluded in a simplified illustration shown above.

3.2 Take-Off Distance Available (TODA)

The declared distance of the **available take-off run length (TORA) plus the clearway** (or any remaining runway beyond the departure end of TORA) is called TODA. That means, in comparison to the TORA, whenever a runway is provided with a CWY, TODA additionally will include the length of that clearway.

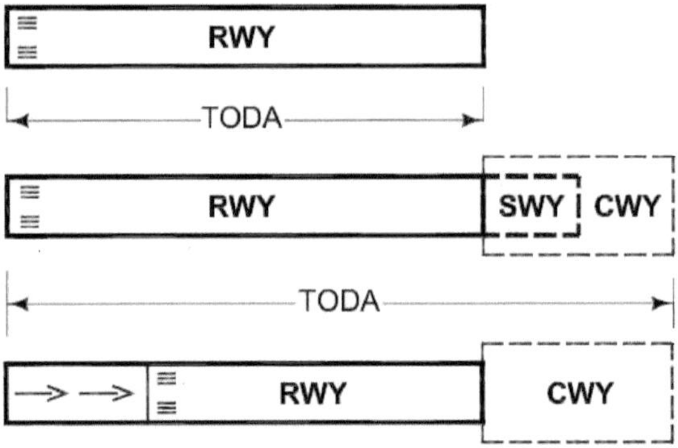

IMAGE 6: SIMPLIFIED ILLUSTRATION OF TODA (*OWN ILLUSTRATION*)

The usable TODA length is controlled by obstacles present in the departure area and aircraft performance. With the provision of a clearway (which is nearly clear of any obstacles) a **safety area** beyond the end of runway is established, which gives the aircraft enough space to perform a secure initial take-off climb. Nevertheless, the useable TODA length shall be determined by the aircraft operator before each take-off and requires knowledge of each controlling obstacle in the departure area (*FAA 2012, p. 64*).

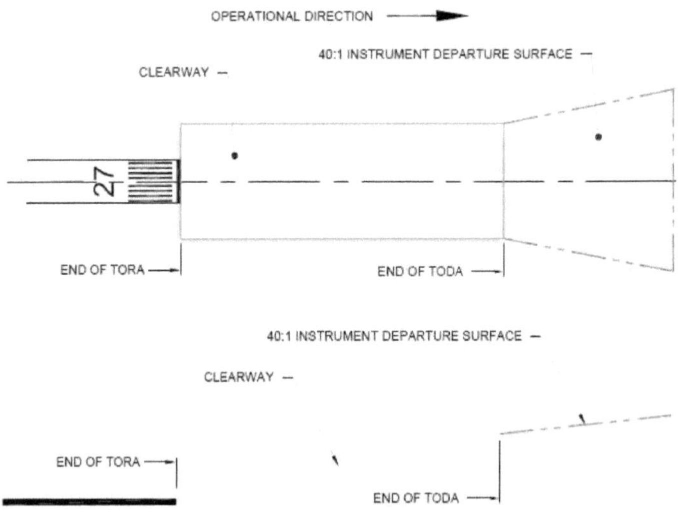

IMAGE 7: EXTENDED TODA BY USE OF A CLEARWAY (*FAA 2012, P. 64*)

3.3 Accelerate Stop Distance Available (ASDA)

According to ICAO Annex 14, ASDA is defined as the **length of the take-off run available (TORA) plus the length of the stopway**, if provided. In other words a pilot can compute an accelerate/stop distance by summing the available runway length and the stopway length.

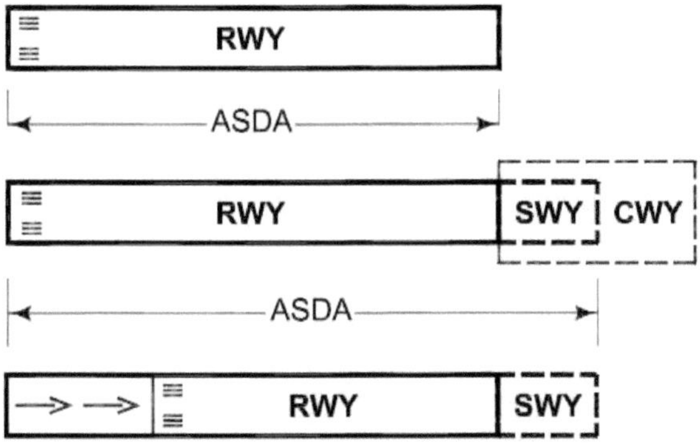

IMAGE 8: SIMPLIFIED ILLUSTRATION OF ASDA (*OWN ILLUSTRATION*)

This calculation, along with other factors, is used by the pilot to determine aircraft loading and performance requirements. In cases of aborting the take-off below the speed of V_1 the pilot has more in his hands than only the dedicated runway. In contrast to the TORA he may also use the stopway to decelerate the aircraft safely. The provision of a stopway by the airport operator as component of usable runway length shall be considered by pilots, when planning the take-off for certain airplanes depending on their certification basis. Because most modern jet transport category airplanes **can take advantage** of the designated SWY, when calculating maximum allowable takeoff weight and performance speeds, as limited by accelerate-stop distance requirements. It simply means that the provision of SWY **allows a much higher take-off mass and faster V_1** as the accelerate stop distance is enlarged. Without any stopway the equation ASDA=TORA is valid (*FAA 2012, p. 65 f.; FAA 2008, p. 1; FAA 2000, p. 1*).

3.4 Landing Distance Available (LDA)

As we know the required landing distance depends on the aircraft performance e.g. landing distance and speed is calculated by the pilots using printed tables or a computer. A flight crew approaching an airport has to make sure that the dimension of the approaching runway is suitable for their aircraft under certain conditions. Assistance in flight pre-planning provides the published Landing Distance Available (LDA). LDA is defined as the length of the runway which is declared available and suitable for the **ground run of an aircraft landing**. The LDA commences at the threshold.

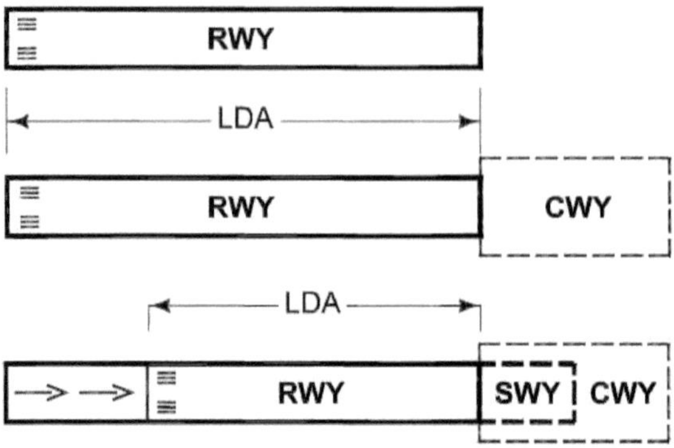

IMAGE 9: SIMPLIFIED ILLUSTRATION OF LDA (*OWN ILLUSTRATION*)

Where a runway has a displaced threshold, the LDA will be reduced by the distance the threshold is displaced, as shown in Image 9. A displaced threshold affects only the LDA for approaches made to that threshold; all declared distances for operations in the reciprocal direction are unaffected. Any portion of a SWY/CWY is not included in the LDA and is unavailable for LDA computations of the flight crew.

To make this complex topic more understandable the following image depicts a combination of all declared distances within a single illustration.

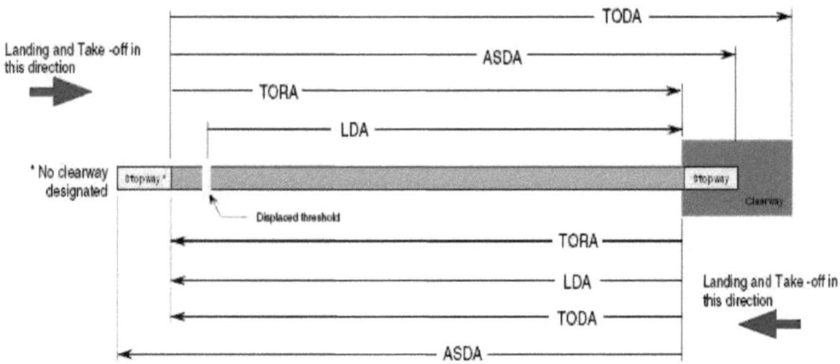

IMAGE 10: DETERMINATION OF DECLARED DISTANCES FOR RUNWAYS (*DEP. OF TRANSPORT CANADA 2011*)

4 Application and Publication of Declared Distances

As already mentioned the statement and specification of declared distances result from the provision of CWY/SWY and/or Displaced Thresholdes. Their application allows airport operators to enhance critical safety factors and increases the allowable aircraft operating parameters without increasing the length of the runway.

But in many cases, especially at large international airports, aircraft/airport operators don't want to use the entire available runway length. At this point a much more important area of application of declared distances appears; the so called **Intersection Takeoffs**. In order to enhance airport capacities, reduce taxiing distances, minimize departure delays, and provide for more efficient movement of air traffic, controllers may initiate intersection takeoffs as well as approve them when the pilot requests. The advantages of an Intersection Takeoff are perfectly obvious. All the reasons mentioned above **finally save money and time**. But reduction of costs and delay must not occur at the expense of safety. And this is exactly the point where the provision of declared distances plays an important role.

Pilots are expected to assess the suitability of an intersection for use at takeoff during their preflight planning. They must consider the resultant length reduction to the published runway length from the intersection intended to be used for takeoff. In those cases the published declared distances for each intersection **support the pilot during pre-planning** the flight. The distances provided in the charts simply tell the pilot: "**If you intent to start your takeoff run out of this intersection, will it fit or not!?**". That means before the intersection can be accepted for takeoff, the minimum runway required for takeoff (under certain circumstances) and the reduced runway length must be taken into consideration. In this context the field layout of "Zweibrücken Airport" is shown in Appendix 1 and acts as an example for providing different values of declared distances for different intersections.

If for any reason a pilot prefers to use a different intersection than the full length of the runway or desires to obtain a certain distance between the intersection and the runway end, the pilot is expected to inform Air Traffic Control (ATC) accordingly. Controllers will issue the measured distance from the intersection to the runway end to any pilot who intends o perform this kind of takeoff. For all military aircraft ATC is obliged to transmit the necessary values. Furthermore, airport operators are encouraged by Eurocontrol to display the relevant TORA value information additionally on the taxi guidance signs or holding points used for intersection takeoffs (*Eurocontrol 2013, p. 33*).

To conclude this chapter I will briefly refer to the regulations concerning the publication of declared distances. ICAO Annex 15 (Aeronautical Information Services) clearly calls for the reporting of declared distances, including clearway and stopway length if provided, for each runway direction in the AIP of the respective country by the airport operator. Declared distances shall be depicted on aeronautical charts and related publications or be available in electronic navigation databases for all international airports, even if the distances are simply equal to the runway lengths in both directions. With this, the relevant information are assessable for aircrews at any time and can be used for flight crews in the pre-planning of the takeoff and landing phase (*FAA 2012, p. 70 f.; ICAO Doc 9357, 3.4.2*). With the provision of declared distances the so called **"end customer" highly profits from enhanced take-off and landing capabilities with all positive consequences** on aircrafts' performance depending on their certification basis (e.g. calculating maximum operating weights and/or operating capability including safety factors).
Should the runway declared distances be temporarily reduced for any reason, e.g. during maintenance or construction work, the position and nature of the markings, signs and lighting should be carefully planned to ensure the correct temporary information is displayed during any changes. These reduced runway lengths shall also be carefully communicated to flight crew by NOTAM or AIP entry and to ATSP for inclusion in ATIS, flight briefing material or live radio communication (*Eurocontrol 2013, p. 34*). In any case it is important that the designation of e.g. SWY/CWY and declared distances for runways are **properly documented, accurately reported, and charted on flight documents** intended for use by the pilot or operator (*FAA 2008, p. 1; FAA 2012, p. 69 f.*).

A typical example of the publication of declared distances in combination of a SWY and CWY is abstracted from the Military AIP Germany and depicted in Appendix 2.

5 Runway End Safety Area (RESA)

As mentioned in the introduction the majority of world's jet aircraft accidents events occur during the takeoff and landing phases of flight. Landing overruns, landing undershoots, takeoff overruns, and crashes after takeoff belong to the group of most frequently reported accident types in the world. Records show that while most accidents occur within the boundaries of the runway strip, most fatalities occur near but off the airport area (*Caves 1996, p. 224*). Therefore Annex 14 to the Chicago Convention, ICAO lays down a part of the aerodrome design section its Standards and Recommendations for aerodrome safety areas in particular for a **Runway Strip** and **Runway End Safety Areas (RESA)**.

The **Runway Strip** can be described as a defined area including the runway and stopway, if provided, intended to reduce the risk of damage to aircraft running off a runway and to protect aircraft flying over it during take-off or landing operations. It extends laterally to a specific distance from the runway centerline and provides an area clear of objects which may endanger aircraft (*ICAO Doc 9157, 5.3*). The statistic below clearly shows that beside the risk of veering of left or right the runway there is a high chance that airplanes undershoot or overrun the runway during landing and take-off.

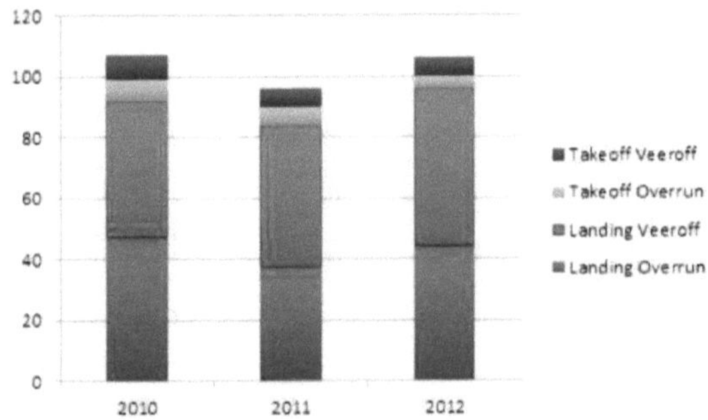

IMAGE 11: RUNWAY EXCURSIONS SINCE 2010 ONWARDS (*AIR TRANSPORT NEWS 2013, P. 13*)

In addition to the Runway Strip requirements, Annex 14 demands a **Runway End Safety Area** especially for those events. Furthermore, ICAO statistics say that particularly aircraft overrunning or overshooting the end of runway suffer significant damage. To minimize such damage, it is considered necessary to provide an additional **safety area beyond the ends of the runway strip**. These areas, known as RESA, should be capable of adequately supporting any aircraft which overruns or undershoots the runway and should be clear of all equipment and installations which are not frangible (*ICAO Doc 9157, 5.3*). ICAO Annex 14 provides the following definition for RESA: "*An area symmetrical about the extended runway centre line and adjacent to the end of the strip primarily intended to reduce the risk of damage to an aeroplane undershooting or overrunning the runway*".

As with all facilities on an airport, RESA is subject to clearly defined regulations. In comparison to the provision of CWY/SWYs or displaced thresholds, the provision of a RESA is mandatory to airport operators. Annex 14 states that at code 3&4 runways (field length of 1200m and more) a RESA which extends a minimum of 90m beyond the end of the runway strip and twice the width of the runway is established. ICAO goes on to recommend that RESAs extend 240m at code 3&4 runways and 120m at code 1&2 runways (field length less than 1200m) and equal to the graded portion of the runway strip are established (*ICAO Annex 14, 3.5*). The graphic below gives an impression of the dimensions postulated by ICAO.

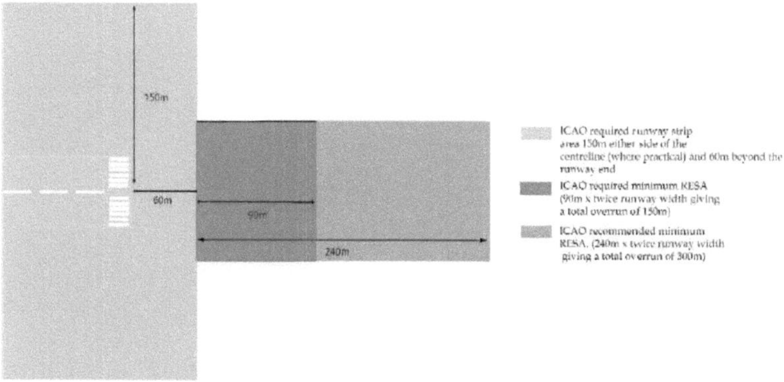

IMAGE 12: RESA DIMENSIONS CODE 3&4 RUNWAYS (*IFALPA 2009, P. 19*)

These standard dimensions have been increased from time to time to address higher safety expectations of aviation users and accommodate current aircraft performance. But applying the new standards to existing airports can be problematic in some cases. Many runways do not meet current standards because they were constructed to an earlier standard (*ACRP 2008, p. 2*). However the provision and proper maintenance of **RESAs minimizes the risk and impact of incidents** relating to undershooting or overrunning of runway, escalating into major accidents, causing serious damages to aircraft and injury to passengers and crew.

6 Conclusion

In my opinion the preceding statement can coincidentally act as the conclusion for the entire project paper. Based on a more or less theoretical introduction it was clearly shown how much importance is attached to safety areas and declared distances in the daily aviation business. As mentioned there has been a noticeable reduction in the number of non-fatal and fatal accidents involving the worldwide commercial jet aircraft fleet over the last decade (*Taylor/Hughes/Godley 2009, p. 1*). However runway excursions are an all too common occurrence in commercial aviation. In the meanwhile it was found out that more than 450 different factors can be judged to be instrumental in the causal chain of events leading to aircraft overruns and undershoots (*van Es 2010, p. 6*). In reality, operations are carried out under varying levels of safety and the provision of safety areas by the airport operators and the publication of declared distances **certainly not prevent those situations completely!** Nonetheless, they **make an essential contribution to minimize the risk** of significant damage and reduce potentially hazards to aircraft and to protect passengers and people on the ground in case of an aircraft veers off or overruns the runway. It is important to recognize that the risk of a runway excursion accident is ever present and that a range of safety measures should be utilized by aircraft operators, and airport owners and managers to ensure **the risk remains at an acceptable level** (*Taylor/Hughes/Godley 2009, p. 78*).

As a rule, every pilot should keep in mind that the runway is only as long as it's most limiting dimension and all performance numbers and aircraft takeoff weight is based on the most limiting runway dimension.

7 Appendix 1

INTERSECTION TAKE-OFF

RWY	TWY	TORA	TODA	ASDA
03	E	2400	2725	2400
03	D	1518	1843	1518
03	C	797	1122	797
21	B	2304	2679	2304
21	C	1901	2276	1901
21	D	1172	1547	1172

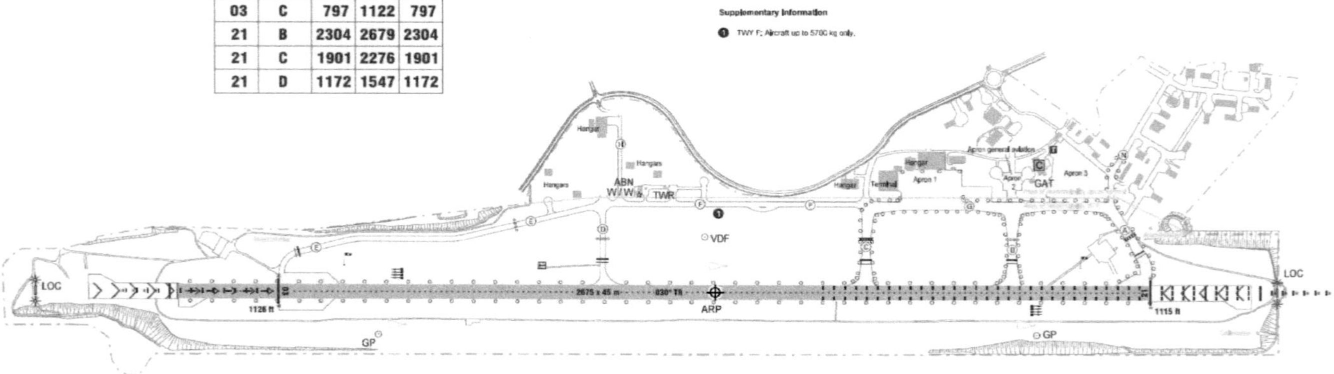

IMAGE 13: AERODROME CHART – ICAO ZWEIBRÜCKEN EDRZ (*AIP GERMANY 2013*)

8 Appendix 2

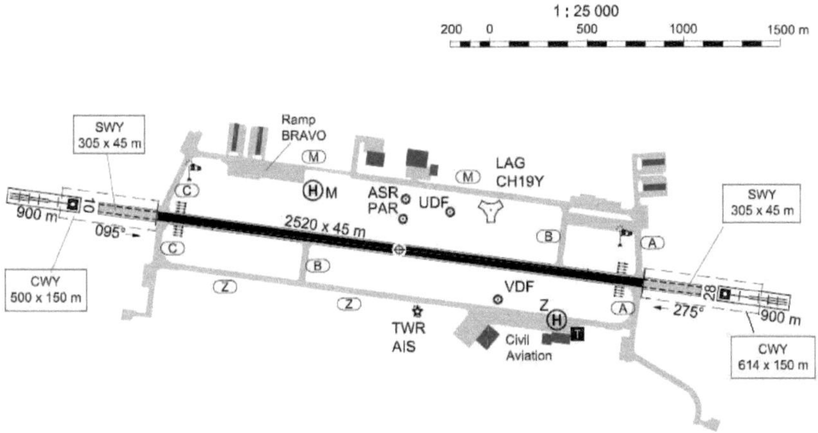

ETNL AD 2.13 Declared Distances

RWY Designator	TORA (m)	TODA (m)	ASDA (m)	LDA (m)	Remarks
1	2	3	4	5	6
10	2520 (2720*)	3134 (3334*)	2825	2520	* 200 m SWY avilable for TKOF O/R via backtrack (PCN 66)
28	2520 (2720*)	3020 (3220*)	2825	2520	

IMAGE 14: AERODROME CHART – LAAGE ETNL (*MIL AIP GERMANY 2012*)

9 Appendix 3

9.1 List of Abbreviations

Abbreviation	Explanation
ADREP	IACO Aircraft Accident/Incident Data Reports
ACRP	Airport Cooperative Research Program
ASDA	Accelerate Stop Distance
AIP	Aeronautical Information Publication
ATC	Air Traffic Control
CWY	Clearway
FAA	Federal Aviation Administration
ICAO	International Civil Aviation Organization
LDA	Landing Distance Available
RESA	Runway End Safety Area
SWY	Stopway
TODA	Take-Off Distance Available
TORA	Take-Off Run Available
V_1	Takeoff Decision Speed

TABLE 1: LIST OF ABBREVIATION

9.2 Bibliography

Air Transport News (2013): *Safety Report 2013: An Exceptional Year in Global Aviation Safety*, downloaded from "http://www.atn.aero/analysis.pl?acateg=survey" on 13.05.13

Ashford, N. J. / Mumayiz, S. A. / Wright P. H. (2011): *Airport Engineering: Planning, Design, and Development of 21st-Century Airports,* 4th Edition, New Jersey

ACRP (2008): *Analysis of Aircraft Overruns and Undershoots for Runway Safety Areas*, downloaded from http://onlinepubs.trb.org/onlinepubs/acrp/acrp_rpt_003.pdf on 30.05.13

Bos, T. (2013): *Takeoff Performance Margins*, downloaded from "https://www.vnv.nl/commissiepub/4266" on 23.05.13

Caves, R. (1996): *Control of Risk Near Airports*, published in: Built Environment, Vol. 22, No. 3, pp. 223-233

Croucher, P. A. (2007): *JAR Professional Pilot Studies*, 1st Edition, United Kingdom

Department of Transport Canada (2011): *Aerodrome Dimensions and Related Information*, downloaded from "http://www.tc.gc.ca/media/images/ca-publications/figure2-1.gif" on 26.05.13

Eurocontrol (2013): *European Action Plan for the Prevention of Runway Excursions, Edition 1.0*, downloaded from "http://www.eurocontrol.int/documents/european-action-plan-prevention-runway-excursions" on 08.06.13

FAA (1999): *Runway Protection Zone and Airport Object Clearing Policy*, downloaded from http://www.faa.gov/airports/great_lakes/airports_resources/ppms/media/5300-1B.pdf on 20.05.13

FAA (2000): *Use of the Term Stopway*, downloaded from "http://www.faa.gov/airports/airport_safety/certalerts/media/cert0003.pdf" on 20.05.13

FAA (2004): *Advisory Circular 150-5300-13, Change 8 to Airport Design*, downloaded from "www.faa.gov/documentLibrary/media/Advisory_Circular/150_5300_13_chg8.doc" on 24.05.13

FAA (2009): *Advisory Circular 150-5300-18B, General Guidance and Specification for Submission of Aeronautical Surveys to NGS*, downloaded from "http://www.faa.gov/documentLibrary/media/Advisory_Circular/150_5300_18b.pdf" on 23.05.13
FAA (2012): *Advisory Circular 150-5300-13A, Airport Design*, downloaded from "http://www.faa.gov/documentLibrary/media/Advisory_Circular/draft_150_5300_13a.pdf" on 23.05.13
IFALPA (2009): *Runway Safety Manual*, downloaded from "http://www.ifalpa.org/downloads/Level1/Briefing%20Leaflets/Airport%20Issues/Runway%20Safety%20Manual%201.2.pdf" on 08.06.13
LPS Aviation (2007): *Airport Runway, Optimal Lengths and Issues*, downloaded from "http://www.dot.gov.nt.ca/_live/documents/content/FinalGNWTRunwaysCompleteWeb.pdf" on 23.05.13
Taylor, R. P. / Hughes, K. /Godley, S. (2009): *Runway Excursions, Part 1: A worldwide review of commercial jet aircraft runway excursions*, downloaded from "http://www.atsb.gov.au/publications/2009/AR2008018_1.aspx" on 08.06.13
van Es, G. W. H. (2010): *A Study of Runway Excursions from a European Perspective*, downloaded from "http://www.nlr-atsi.nl/downloads/a-study-of-runway-excursions-from-a-european-p.pdf" on 08.06.13

TABLE 2: BIBLIOGRAPHY

9.3 Images

9.4 Tables